Did you enjoy this issue of BioCoder?

Sign up and we'll deliver future issues and news about the community for FREE.

http://oreilly.com/go/biocoder-news

BioCoder #9

OCTOBER 2015

Beijing • Boston • Farnham • Sebastopol • Tokyo

BioCoder #9

Printed in the United States of America.

Published by O'Reilly Media, Inc., 1005 Gravenstein Highway North, Sebastopol, CA 95472.

O'Reilly books may be purchased for educational, business, or sales promotional use. Online editions are also available for most titles (*http://safaribooksonline.com*). For more information, contact our corporate/institutional sales department: 800-998-9938 or *corporate@oreilly.com*.

Editors: Mike Loukides and Nina DiPrimio
Production Editor: Nicole Shelby
Copyeditor: Kim Cofer
Proofreader: Marta Justak
Interior Designer: David Futato
Cover Designer: Randy Comer
Illustrator: Rebecca Demarest

October 2015: First Edition

Revision History for the First Edition

2015-10-13: First Release

978-1-491-93096-0

[LSI]

Contents

Foreword

Mike Loukides

In past issues of *BioCoder*, we've published about the problem of reproducibility in science and its connection to lab automation and the revolution in software tools for the life sciences.

In *BioCoder #9* (*http://www.oreilly.com/biocoder*), these themes come around again in exciting new ways. I've always thought of lab robots in terms of fluid handling, but Dave Zucker takes robotics into a different direction: can you automate the handling and classification of fruit flies, the raw material of many genetics experiment? The FlySorter (*http://www.flysorter.com/*) represents both clever mechanical design and a collaboration with computer vision experts. It's a work in progress that I want to watch very closely.

Edward Perello writes about *Putting the Tech in Biotech*. We have many tools for design and engineering in synthetic biology. The problem is that they aren't good tools. They're tools written by biologists, who understand the biological issues, but don't understand how to build tools that are intuitive and easy to use. While we've all used software that's neither pleasant nor intuitive, the problem is almost always that the coders never talked to the users and never understood the problems they were trying to solve. Perello suggests ways to bridge the gap among the programmers, designers, and biologists: these collaborations, possibly like the team that was building the FlySorter, allow programmers and computer scientists to make important contributions without requiring them to become biologists.

When they discuss the problem of describing protocols precisely and unambiguously, Ben Miles and Tali Herzka are also talking about the connection between software and biology. We are developing standards, such as Autoprotocol (*http://autoprotocol.org/*), for describing protocols in ways that can be implemented by robots. We also need standards for the way humans communicate: what do we mean when we use terms like "mix," which are intuitive, convenient, and imprecise? In their article, Ben and Tali show Wet Lab Accelerator (*http://bit.ly/wet-lab-acc*), a protocol design tool by Autodesk that lets biologists drag and drop steps onto a worksheet, and then export the result in Autoprotocol. Again, skilled software designers are collaborating with biologists to produce a new generation of tools.

Tim Gardner talks about another collaboration between computer scientists and biologists. Tim's company, Riffyn (*http://riffyn.com/*), has teamed up with PLOS (*http://plog.org/*) to build an open source tool that builds vocabularies by mining PLOS publications. The first phase of ResourceMiner (*https://www.mozillascience.org/projects/resourceminer*) was completed at the Mozilla Science Lab's Global Sprint (*https://www.mozillascience.org/global-sprint-2015*), which brings software developers together with scientists to hack on open science projects.

Our current tools, awkward as they are, have gotten the biological revolution started. To win, we'll need tools that are easier to use and more powerful. Those tools are appearing—and more are on the horizon, as software developers team up with biologists to build our future.

Afineur

Glen Martin

In the realm of coffee, kopi luwak (*http://www.most-expensive.coffee/*) is the ne plus ultra, the best of the very best, a coffee that transcends superlatives. At least, that's what people who have tasted it claim. Admittedly, that's a rather small and select cohort, due to the fact that kopi luwak sells for up to $600 a pound; a single demitasse of the revered brew can cost $90. Why so much lucre for a simple cup of joe? Three words: Indonesian civet cats. (*http://bit.ly/civet-pics*)

To make kopi luwak, you need these small and rather adorable omnivores as much as you need coffee beans. In their native Indonesia, civet cats roam freely through the nation's vast coffee plantations, gobbling the ripening "cherries" directly from the trees. They digest the flesh of the cherries readily enough, but the seeds—the coffee beans—pass through intact. And yet, they are not the same beans that hung on the tree prior to ingestion. Not surprisingly, considering the dynamic environment of a typical civet gut, their flavor and aroma profiles have been transformed.

Somewhere, at a time lost in the misty annals of coffee culture, an Indonesian plantation worker looked down at a pile of civet excrement brimming with undigested coffee beans and wondered: Could you brew that? That moment evokes Jonathan Swift's rumination on certain shellfish, *viz.*, "...It was a bold man that first ate an oyster." In any event, a gelatinous and gelid raw oyster ultimately slid down the gullet of an adventurous human, who found it good. And at a particular point, somebody picked out the beans from a clump of poop produced by a highly caffeinated civet cat, cleaned them, roasted them, brewed them, and hailed the resulting beverage as ambrosial. It was determined that civet coffee was far less bitter than regular coffee, evincing an abundance of disparate and appetizing overtones. It was, in short, delicious.

And yet—let's face it—the idea of getting your coffee from the digestive tract of a civet cat is rather unsavory (though it also admittedly has a certain madcap marketing appeal). Too, the supply perforce is limited: there are only so many civet cats around that you can harness for coffee production. Moreover, animal cruelty issues are involved. Civet cat farms (*http://bit.ly/civet-cat-cruelty*) in Indonesia, the Philippines, and Vietnam—where the animals are housed in batteries of cages and fed a steady diet of coffee beans—have been slammed by animal rights groups.

So how do you get the supernal flavors of civet cat coffee without civet cats? Enter Booklyn-based Afineur (*http://www.afineur.com/*), producers of "cultured" coffee. Kopi Luwak isn't really about civet cats, says Camille Delebecque (*http://bit.ly/delebecque*), the company's co-founder and CEO. It's about fermentation: the process that occurs in civet cat intestines. Duplicate the fermentation process in a controlled environment, and you should be able to produce kopi luwak *sans* civet cats.

Indeed, says Delebecque, who took his PhD in synthetic biology at Harvard, that's just what Afineur has done. The company uses standard biofermenters to transmute green coffee beans prior to roasting.

"We were very successful with the first coffee we produced in greatly reducing astringency and bitterness," says Delebecque. "We found that those two flavors tend to mask other, more interesting flavors very effectively. Once we were able to reduce them, we found all these floral notes and fruity qualities coming through."

Along with minimizing bitterness and enhancing desirable flavors, says Delebecque, Afineur's proprietary fermentation process also improves the nutritional value of coffee. "Fermentation increases the vitamin B and antioxidant content," Delebecque says.

Afineur employs a variety of microorganisms collected from different sources, says Delebecque; none have been genetically modified.

"We haven't manipulated any microbes directly, but the microbiological communities we establish are completely synthetic," he says. "You wouldn't find them in nature. We adjust the ratios (of various microbes) depending on the flavors we're trying to emphasize."

Afineur's staffers aren't just thinking of fermentation-induced flavors as they fine-tune their microbial communities. They also have the flavors and aromas associated with the Maillard reaction (*http://bit.ly/maillard-reaction*) in mind: the reduced, caramelized qualities created by toasting. They work with green beans in the biofermenters, but they are always aware that the final product will come out of a roaster.

"I think the symphonic analogy is a good one to use in describing our work," Delebecque says. "We think we can produce a whole array of profiles in our coffee,depending on how we 'conduct' our creative microbes."

Taste, of course, is—well, a matter of taste. So how do you maintain quality control with a cup of coffee? Delebecque combines two approaches: gas chromatography and organoleptic testing (i.e., the employment of the human senses). Gas chromatography pinpoints the precise ration of volatile organic compounds in specific flavor profiles, and the human nose, palate, and tongue determine whether those flavors taste good.

Delebecque acknowledges that it's no coincidence that he and Afineur's co-founder, Sophie Deterre (*http://bit.ly/deterre*), are French. It's not just the Gallic love and respect for good food that drove their interest in making a better cup of coffee, he says; it's also the fact that many of the iconic foods and beverages associated with French cuisine are predicated on fermentation.

"Fermentation is essential to our cheese, wines, and beer," he observes. "So it's really about culture. Like all French men and women, I grew up on fermented foods. They're part of who we are, essential to what we value in life."

In fact, it's likely that Afineur will move beyond coffee. There's a whole world of comestibles out there, Delebecque believes, that can be improved by his stable of creative microbes.

"We think tailored fermentation can be applied widely," Delebecque says. "We're setting up a platform technology that will ultimately enhance the flavor and nutrition of many foods."

The Afineur gospel, certainly, seems to have wide appeal. The company recently launched a Kickstarter (*http://bit.ly/cultured-kickstarter*) campaign. As of this writing, with 20 days to go, Delebecque and crew have received pledges of $41,301 from 716 backers: $26,301 above their $15,000 goal.

Glen Martin *covered science and environment for the San Francisco Chronicle for 17 years. His work has appeared in more than 50 magazines and online outlets, including Discover, Science Digest, Wired, The Utne Reader, The Huffington Post, National Wildlife, Audubon, Outside, Men's Journal, and Reader's Digest. His latest book, Game Changer: Animal Rights and the Fate of Africa's Wildlife, was published in 2012 by the University of California Press.*

Marketing Synthetic Biology

Karen Ingram

Perhaps the most well-known marketing professional is creative director Don Draper, the main character of the AMC drama *Mad Men* (*http://bit.ly/amc-mad-men*). Throughout the series, Don is the creator and custodian of the "Big Idea," spinning narratives and captivating clients and consumers alike. The series shows Don wooing clients and gallivanting with vixens, but it also shows him attending French cinema, hanging with beatniks, and participating in other cultural experiences relevant for the time period. Throughout the series, Don explores cultural context; inspiration in which to nestle his product or service, a catalyst for his "Big Idea." In the grand finale of *Mad Men*, Don Draper is a creative director at McCann Erikson struggling to find himself, or maybe just a metaphor for a big ticket client. Through a series of experiences at a picturesque ocean side retreat in California, Don reaches his sought-after epiphany—he reappropriates the hand holding, hair swinging, utopian culture of the early 70s hippie enlightenment into a Coke ad. You know—THAT Coke ad (*http://bit.ly/coke-ad-1971*): "It's the Real Thing." It just so happens that McCann Erikson, a real agency in New York, was indeed responsible for that ad... the real thing.

Don Draper is a product of traditional marketing (advertising of the "genius steals" nature). Like Don Draper, I was a creative director at McCann Erikson. I was also a creative director at an agency called Campfire. Campfire is a newer marketing model, that of storytelling and social engagement, which has sprung from rapid evolutions in technology and media.

Campfire started when Mike Monello and Greg Hale—two producers of the genre blurring indie horror film, *The Blair Witch Project* (1999 (*http://bit.ly/blair-witch*))—opened up an agency inspired by the tactics they'd used for their own indie film. *The Blair Witch Project* is a fictional narrative that takes the shape of a documentary. The plot revolves around footage found of a missing film crew in Burkittsville, Maryland, who had been investigating the myth of the Blair Witch. The film was first introduced to the public on the Internet via a website outlining the folklore around the Blair Witch. Nevermind that the tale of the Blair Witch was a bit of folklore that no one had ever heard of. The website sparked curiosity and engagement, and by the time the film was released in the cinema, people were excited to see it. They were "in on it," whether they believed the footage was real or not. People asked questions like: Who (or what) is the Blair Witch? What happened to the film crew? How was the footage found? Did these events happen?

Campfire has used this methodology—technology and narrative—to tap into what a Fast Company article (*http://bit.ly/fc-rabbit-hole*) refers to as "the curiosity gap, " and it has worked.

In its campaigns and storytelling, Campfire seeds the breadcrumbs that drive engagement and generate memes. They allow for fans to invest in the narrative, and even build the universe around the story architecture they create. Through access to production and creative tools, fans of products, services, and even TV shows like *Mad Men* now create art, media, and stories around things they feel passionately about.

There is a way that storytellers of all types map out different levels of public engagement. In a narrative, it's the "hero's journey." In digital marketing, the analogue is the "user journey." The user journey is the path a consumer or user takes to engage with a product or service. Depending on the product or service, there are several ways to map it out, but most user journeys look like this:

Awareness
: The user may have heard of the product/service; a cursory knowledge, but doesn't have a connection to it.

Investigation
: The user has a full awareness and comprehension of the product/service.

Acceptance
: The user has explored the product or service and developed the opinion that it is of interest to her. Perhaps she has purchased the product or service.

Retention
: The user is actively engaged, and has achieved competence in usage of the product or service.

Advocacy
: The user is a well-versed ambassador for the product or service.

So does this type of thinking apply to the introduction of a new technology? If it does, how? What do people think about synthetic biology? *Do* they think about it? What associations exist now?

Below are a few vignettes that speak to the application of this line of thinking. It can reveal collective thinking, rather than individual, and how it might apply to synthetic biology; marketing as a means to communicate what may or may not be favorable with new technology.

Awareness

The first time I attended Synberc, I participated in a science communications workshop facilitated by journalist Christie Nicholson of the Alan Alda Center for Communicating Science. Christie shared a "man on the street" video in which members of the "general public" (aka non-scientists) were asked about their knowledge of scientific terms like "nano," "basic research," "GMO," and a few other terms. Most terms didn't register too much with the interviewees. Some, of course, had less than favorable responses, with the exception of "nano."

Perhaps it's worth considering: does "syn" or "synthetic" imply that the origin is artificial, further supporting that synbio is "unnatural?" Has the terminology hit the "euphemism treadmill?" Or can the "synthetic" in synbio maintain its integrity as "combination or composition, in particular," or "the combination of ideas to form a theory or system (Google)."

Investigation

In the "man on the street" interviews, why did the term "nano" register? Most interviewees mentioned the iPod nano. If you google the term "nano," that will be reinforced. Cars, Quadra copters, and other random images are also present–consumer products. On the other hand, if you google "nanotechnology," you'll get an array of cogs, micromachinery, graphene grids, neurons...."GMO" reveals innocent fruits and veggies punctured by syringes, impossible hybrids, franken food, and red circles with slashes. "Synthetic biology" reveals charts and diagrams, mostly academic or industry-specific imagery. This implies that at present, there doesn't seem to be much of a context around synbio, unless it's just to those those who are working directly in the area.

~~Acceptance~~ Engagement

Kevin Hooyman's illustrations are charming and naïve, like a keepsake doodled on a restaurant placemat. Visit his website, and you'll be greeted by a grid of meekly smiling, sloth-monkey-bigfoot characters, centaurs, the occasional tree-dwelling crocodile—clever hand-rendered creative concoctions. For his first assignment for the *New York Times*, he was asked to "capture some aspect of the GMO debate to illustrate letters to the editor in the opinions section" in response to a piece by Mark Lynas (*http://bit.ly/nyt-gmo-food*). Lynas is a researcher at the Cornell Alliance for Science who had recently decided that it was "anti-science" to discount research stating that GMOs are safe, while accepting evidence that climate change is real. Hooyman googled GMOs, like most illustrators do when looking for inspiration. His search yielded "mutant food, but it seemed pretty overdone and blunt and one-sided."

In the end, the *NYT* selected an innocuous image of stalks of strawberries (*http://bit.ly/nyt-when-food*) perched in a beaker from the images that Hooyman produced: no syringes. His engagement yielded a different and more carefully considered perspective. The image depicts the science encouraging life to flourish, rather than the science manipulating and forcing life to behave in a manner that would be deemed "unnatural."

Retention

What does it mean that a technology is being represented as "one-sided?" How can that boundary be traversed to those who may be future stakeholders? Over the past two years or so, I've been involved in the creation of a book about synthetic biology, part of a team amassed by Natalie Kuldell, along with Rachel Bernstein and Katie Hart. *Biobuilder* (*http://bit.ly/biobuilder*) is divided into two parts. The first section of the book goes over fundamental topics: synthetic biology, biodesign, DNA engineering, and bioethics.The second half of the book includes five labs; four wet, labs and one electronic lab. Even if you never walked into a lab, you'd walk away with an informed perspective on synthetic biology.

In an impulsive burst of excitement around the release of the *Biobuilder* book, Natalie, Rachel, Katie, and I began to ask folks who purchased *Biobuilder* to tweet a picture of the book with the hashtag #wheresbiobuilder. Tweets came in from around the world: from labs, offices, kitchen tables... people got creative; high school kids, scientists, even pets made their way into the collection of imagery. The last time I checked, cats couldn't read, but we all know they own the Internet (see Figure 2-1).

Figure 2-1. I can has Biobuilder. I can learn about synthetic biology.

~~Advocacy~~ Creativity

You'll note that I replaced "Advocacy" with "Creativity." The reason why I did this is because creativity allows for more hands-on interaction. This is an important element of getting the public invested in new technology. They need to be immersed and involved in order to see applications, rather than have the applications delivered to them prepackaged. The clarion call isn't "Let's all be genetic engineers," rather it is, "Let's see what this is all about." The best way to do this is by involving people of other disciplines.

Community biotech labs like Genspace (*http://genspace.org/*), Biocurious (*http://biocurious.org/*), La Pailasse (*http://lapaillasse.org/*), Biologik (*http://biologiklabs.org/*), Waag society (*http://waag.org/en*), as well as maker spaces like 757 Makerspace (*http://www.757makerspace.com/*) (Norfolk, VA) and convergent cultural spaces like EMW bookstore (*http://www.emwbookstore.com/*)(Boston, MA) are great places for collaborations. Let's expand that landscape. Encouraging communication between those who see the lure of the "curiosity gap" and practitioners in the field of synthetic biology would serve both communities, resulting in fruitful collaborations that would likely serve all parties.

Any creative director worth his or her weight in salt (or Old Fashioneds, DNA-iquiri (*http://bit.ly/dna-iquiri*)s, whatever) wouldn't touch a creative brief that wasn't backed by a solid brand strategy. Luckily, a quick survey of quotes pulled from the websites of leading groups and institutions focused on synthetic biology read like brand strategies:

LEAP (http://synbioleap.org)
: Catalyzing a community of emerging leaders in synthetic biology to create bold new visions and strategies for developing biotechnology in the public interest.

Synberc (http://www.synberc.org/)
: Synberc is a major U.S. research program to make biology safer and easier to engineer.

SynbiCITE (http://synbicite.com/)
: SynbiCITE is a pioneering Innovation and Knowledge Centre (IKC) dedicated to promoting the adoption and use of synthetic biology by industry.

The Wilson Center: Synthetic Biology Project (http://bit.ly/wilson-syn-bio)
: The Synthetic Biology Project aims to foster informed public and policy discourse concerning the advancement of synthetic biology—an emerging interdisciplinary field that uses advanced science and engineering to make or redesign living organisms, such as bacteria, so they can carry out specific functions.

So it appears as though we've got the perfect primers to build awareness. Let's get creative. If you—whomever you are—are inspired by what you've read above and would like to get involved in collaborations that will help shape an informed context around synthetic biology, please reach out to me at *karen@krening.com*.

Karen Ingram *is a designer, artist and creative director. She is a co-organizer of Brooklyn science cabaret and The Empiricist League, and is a board member of SXSW Interactive. Her work has appeared in many publications including titles from Die Gestalten (Berlin), Scientific American, and The FWA, where she was named a "Digital Pioneer." Ingram is coauthor of a synthetic biology textbook, Biobuilder (O'Reilly, 2015), and as a 2015 Synthetic Biology LEAP (Leadership Excellence Accelerator Program) fellow, Karen is recognized as an emerging leader in synthetic biology. You can learn more about her work at http://www.kareningram.com or follow her on Twitter (@krening).*

Crowdsourcing a Language for the Lab

Marcus Carr and Timothy Gardner

Neither human nor machine communication can happen without language standards. Advancing science and technology demands standards for communication, but also adaptability to enable innovation. ResourceMiner is an open source project attempting to provide both.

From Gutenberg's invention of the printing press to the Internet of today, technology has enabled faster communication, and faster communication has accelerated technology development. Today, we can zip photos from a mountaintop in Switzerland back home to San Francisco with hardly a thought, but that wasn't so trivial just a decade ago. It's not just selfies that are being sent; it's also product designs, manufacturing instructions, and research plans—all of it enabled by invisible technical standards (e.g., TCP/IP) and language standards (e.g., English) that allow machines and people to communicate.

But in the laboratory sciences (life, chemical, material, and other disciplines), communication remains inhibited by practices more akin to the oral traditions of a blacksmith shop than the modern Internet. In a typical academic lab, the reference description of an experiment is the long-form narrative in the "Materials and Methods" section of a paper or a book. Similarly, industry researchers depend on basic text documents in the form of standard operating procedures. In both cases, essential details of the materials and protocol for an experiment are typically written somewhere in a long-forgotten, hard-to-interpret lab notebook (paper or electronic). More typically, details are simply left to the experimenter to remember and to the "lab culture" to retain.

At the dawn of science, when a handful of researchers were working on fundamental questions, this may have been good enough. But nowadays, this archaic method of protocol recordkeeping and sharing is so lacking that half of all biomedical studies are estimated to be irreproducible, wasting $28 billion each year of US government funding.[1] With more than $400 billion invested each year in biological and chemical research globally, the full cost of irreproducible research to the public and private sector worldwide could be staggeringly large.

One of the main sources of this problem is that there is no shared method for communicating unambiguous protocols, no standard vocabulary, and no common design. This makes it harder to share, improve upon, and reuse experimental designs—imagine if a construction company had no blueprints to rely on, only ad hoc written documents describing their project. That's more or less where science is today. It makes it hard, if not impossible, to compare and extend experiments run by different people and at different times. It also leads to unidentified data errors and missed observations of fundamental import.[2]

To address this gap, we set out at Riffyn (*http://www.riffyn.com/*) to give lab researchers the software design tools to communicate their work as effortlessly as sharing a Google doc, and with the precision of computer-aided design. But precise designs are only useful for communication if the underlying vocabulary is broadly understood. It occurred to us that development of such a common vocabulary is an ideal open source project.

To that end, Riffyn teamed up with PLOS (*https://www.plos.org*) to create ResourceMiner (*https://www.mozillascience.org/projects/resourceminer*). ResourceMiner is an open source project to use natural language-processing tools and a crowdsourced stream of updates to create controlled vocabularies that adapt to researchers' experiments and continually incorporate new materials and equipment as they come into use.

A number of outstanding projects have produced, or are producing, standardized vocabularies (BioPortal (*http://bioportal.bioontology.org/*), Allotrope (*http://www.allotrope.org/*), Global Standards Institute (*http://www.gbsi.org/*), Research Resource Initiative (*https://scicrunch.org/resources*)). However, the standards are constantly battling to stay current with the shifting landscape of practical use patterns. Even a standard that is extensible by design needs to be manually extended. ResourceMiner aims to build on the foundations of these projects and extend them by mining the incidental annotation of terminology that occurs in the scientific literature—a sort of indirect crowdsourcing of knowledge.

We completed the first stage of ResourceMiner during the second annual Mozilla Science Global Sprint (*https://www.mozillascience.org/global-sprint-2015*)

1 Freedman LP, Cockburn IM, Simcoe TS (2015). "The Economics of Reproducibility in Preclinical Research." PLoS Biol 13(6): e1002165. doi: 10.1371/journal.pbio.1002165.

2 Gardner, TS (2013). "Synthetic biology: from hype to impact." Trends in Biotechnology, March 31(3), 123-5. doi: 10.1016/j.tibtech.2013.01.018.

in early June. The goal of Global Sprint is to develop tools and lessons to advance open science and scientific communication, and this year's was a rousing success (Mozilla has posted a summary online (*https://www.mozillascience.org/you-did-this-mozilla-science-global-sprint-2015/*)). More than 30 sites around the world participated, including one at Riffyn for our project.

The base vocabulary for ResourceMiner is a collection of ontologies sourced from the National Center for Biomedical Ontology (Bioportal) and Wikipedia. We will soon incorporate ontologies from the Global Standards Initiative and the Research Resource Initiative as well. Our first project within ResourceMiner was to annotate this base vocabulary with usage patterns from subject-area tags available in PLOS publications. Usage patterns will enable more effective term search (akin to PageRank) within software applications.

Out of the full corpus of PLOS publications, about 120,000 papers included a specific methods section. The papers were loaded into a MongoDB instance and indexed on the Methods and Materials section for full-text searches. About 12,000 of 60,000 terms from the base vocabulary were matched to papers based on text-string matches. The parsed papers and term counts can be accessed on our MongoDB server, and instructions on how to do that are in the project GitHub repo (*https://github.com/RiffynInc/ResourceMiner*). We are incorporating the subject-area tags into Riffyn's software to adapt search results to the user's experiment and to nudge researchers into using the same vocabulary terms for the same real-world items. Riffyn's software also provides users with the ability to derive new, more precise terms as needed and then contribute those directly to the ResourceMiner database.

The next steps in the development of ResourceMiner (*and where you can help!*) are to (1) expand the controlled vocabulary of resources by text mining other collections of protocols, (2) apply subject tags to papers and protocols from other repositories based on known terms, and (3) add term counts from these expanded papers and protocols to the database. During the sprint, we identified two machine learning tools that could be useful in these efforts and could be explored further: the Named Entity Recognizer (*http://nlp.stanford.edu/software/CRF-NER.shtml*) within Stanford NLP (*http://nlp.stanford.edu/software/index.shtml*) for new resource identification and Maui (*https://github.com/zelandiya/maui*) for topic identification. Our existing vocabulary and subject area database provide a set of training data.

Special thanks to Rachel Drysdale, Jennifer Lin, and John Chodaki at PLOS for their expertise, suggestions, and data resources; to Bill Mills and Kaitlin Thaney at MozillaScience for their enthusiasm and facilitating the kick-off event for ResourceMiner; and to the contributors to ResourceMiner.

If you'd like to contribute to this project or have ideas for other applications of ResourceMiner, please get in touch! Check out our project GitHub repo (*https://github.com/RiffynInc/ResourceMiner*), in particular the wiki and issues, or contact me at mcarr@riffyn.com.

***Marcus Carr** is a software engineer at Riffyn and an experimental scientist experienced in biophysical chemistry and enzymology. He was previously a research scientist at Kiverdi, a biotech/renewable chemicals start-up. Marcus has conducted research in biochemistry and biophysics at UC Berkeley and Lawrence Berkeley National Lab, and studied chemistry at UC Berkeley and Columbia. Outside of work, he can often be found cycling or (occasionally) running in the Berkeley Hills and beyond.*

***Tim Gardner** is the founder and CEO of Riffyn. He was previously Vice President of Research & Development at Amyris, where he led the engineering of yeast strain and processes technology for large-scale biomanufacturing. He was also an Assistant Professor of Biomedical Engineering at Boston University. Tim has been recognized for his pioneering work in Synthetic Biology by Scientific American, Nature, Technology Review, and the New York Times.*

FlySorter

AN AUTOMATED *DROSOPHILA* CLASSIFICATION SYSTEM

Dave Zucker

I keep a wine fridge in my spare bedroom, and it's full of fruit flies. On purpose.

Figure 4-1. The author's Drosophila storage system

Just a few feet away from these unsuspecting flies is the prototype of a machine designed to efficiently isolate, image, and classify them, all in the name of science.

Drosophila melanogaster—commonly referred to as *fruit flies* or *vinegar flies* —are one of the most frequently used model organisms in biological research (Dietrich et al, 2014 (*http://bit.ly/dietrich-pub-trends*)), in no small part because they are easy to keep and care for. They need little besides an inexpensive food source (which also serves as hydration) and space to grow, and then they multiply quickly. A 12" x 12" tray of small plastic vials can hold thousands of flies, each fly a tiny experimental subject. For under $100 total—for flies, vials, food, and, if you're feeling rich, a $70 wine fridge from Craigslist—anyone can work with *Drosophila.*

Flies were first used to study genetics over 100 years ago, and researchers have developed powerful tools and systems that allow them to learn about disease, development, behavior, neuroscience, and much more, all using the humble fly (Stephenson & Metcalfe, 2013 (*http://bit.ly/stephenson-fruit-fly*)). For all the tools at scientists' disposal, however, there is still a great deal of manual labor involved in fly experiments (Greenspan, 2004 (*http://bit.ly/greenspan04*)). It was a conversation with a friend about fly pushing that pushed this engineer to found a company and develop many prototypes to automatically sort fruit flies by sex, eye color, and other physical characteristics.

Why Sort?

Researchers have isolated thousands of different mutant strains of *Drosophila* as a way to investigate aspects of genetics, physiology, development, and other biological systems. Because most of these mutations are not readily apparent just looking at a fly, scientists have developed a method of linking other genes—ones that code for an externally visible characteristic like eye color, wing shape, production of a fluorescent protein, etc. (see Figure 4-2)—to the ones they're interested in studying. This way, they can more easily distinguish flies that carry the mutation of interest from flies that do not.

Figure 4-2. Different fly phenotypes, from left: wild-type female, wild-type male, white eyes, curly wings, yellow body with white eyes

Labs also frequently isolate males and "virgin" females to set up a mating cross or to tally results at the end of an experiment.

Most labs use a similar approach for sorting:

- Flies are anesthetized with carbon dioxide gas, and the vial is emptied onto a small platform. Exposure to CO_2 immobilizes *Drosophila* temporarily; they recover rapidly once returned to a normal atmosphere.[1]
- The platform—called a *flypad*—is placed under a stereomicroscope, where flies are examined, one at a time.
- Researchers move flies into piles on the pad according to physical characteristics, using a soft-bristled brush to avoid injury.
- Finally, the different piles of flies are transferred (either with the brush or with a mouth pipette) to appropriate destinations: different vials, various experimental assays, etc.

Though the process is relatively simple, it can be time consuming, especially when experiments include tens or hundreds of thousands of flies. I saw an opportunity to free up researchers' time for less menial labor and to solve an interesting engineering problem while improving throughput and accuracy. I decided to automate this process.

Is This Even Possible?

Addressing feasibility early on in a project is good practice. You don't have to build a fully working prototype to know whether something is doable, and more often than not, you *shouldn't*—it's a waste of time, energy, and/or money. I took a few different approaches.

First, I chatted with a number of computer vision and machine learning authorities to learn about imaging and classification technologies that might be helpful. Opinions ranged on the right algorithmic approach, but most agreed it was doable.

And as luck would have it, my brother is one of those experts. Matt Zucker is an Assistant Professor of Engineering at Swarthmore College (*http://bit.ly/matt-zucker*), and studied computer vision and machine learning extensively in gradu-

1 Prolonged exposure can be damaging or lethal, and even with shorter exposures, some effects can linger. For example, CO_2 can alter a fly's sense of taste and smell for hours. Some labs expose flies to cold temperatures as an alternative to CO_2, to avoid these side effects.

ate school. Not only was I able to call him up with questions, but I later persuaded him to join FlySorter as a co-founder.

I built a number of prototype mechanisms in an effort to repeatably image and classify flies. I anesthetized and sent them through a vibrating hopper, hoping they would all come out in a particular orientation. I used air pumps to shoot them through tubes. I took apart web cameras to make cheap, high-resolution microscopes. I illuminated flies with infrared and UV light to try to find highly visible markers. And so on.

Finally, instead of trying to build a fully automatic system, I captured and manually classified hundreds of images of flies by sex, vial after vial, and trained machine learning software on the data. My first datasets yielded a 75%–80% success rate. While hardly good enough for a final product, the rate was very promising, considering how rough the images were.

So a few months in, I felt good. I had proved that machine learning software could differentiate between pictures of anesthetized male and female flies, and I had in mind a robotic platform that would capture images of flies on a pad and then use air pressure to sort them into vials, once software had done the classification.

Making Stuff

I've always been a maker and tinkerer, but to produce something of this complexity required more than just buying parts from McMaster-Carr or Digi-Key. Having used 3D printers extensively at previous jobs, I knew how powerful they could be as a prototyping tool. I did my research, and I bought the parts to build a Kossel Mini (Rocholl, 2012 (*http://reprap.org/wiki/Kossel*)).

The printer was more than just a way for me to make plastic parts. The Kossel Mini is an open source printer, meaning the design is freely available for others to use and modify. It is a delta robot, a platform also well suited for moving a camera and fly manipulator quickly.

I replaced the plastic extruding head on the printer with one that held a high-resolution digital camera and a set of tubes connected to vacuum pumps, and replaced the print bed with a flypad. After some months of tinkering, I had a machine that could image flies on a flypad automatically, and subsequently move individuals to one of three vials atop the machine, using vacuum pumps. Thus, the first version of the sorter was born (see Figure 4-3).

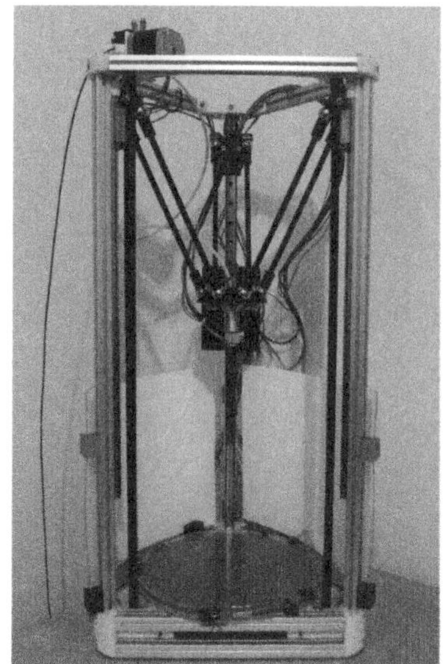

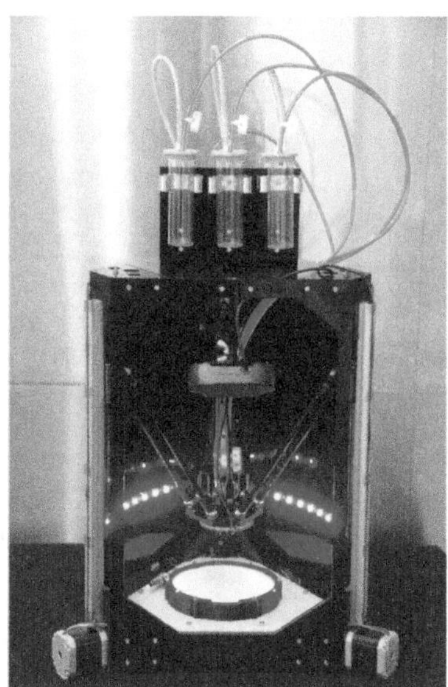

Figure 4-3. Left: Kossel Mini 3D printer; right: FlySorter DCS, version 1.0

Because we were imaging anesthetized organisms, the flies were oriented more or less randomly on the pad: on their backs, on their sides, on their bellies, or legs all akimbo. To account for this inconsistency, we developed a two-step classification process. First, we used a texture classification technique to identify fly body parts in the image (Varma & Zisserman, 2005 (*http://bit.ly/zisserman-texture*)).

Using the labeled image, we could then orient the image and narrow our analysis to a region of interest. For example, to classify by sex, we rotated each image so the head pointed up and then cropped out everything but the abdomen (where the most significant differences between males and females would be obvious). We passed that standardized image to a second classifier to determine the sex of the fly (Shan, 2012 (*http://bit.ly/shan-local-binary*)). See Figure 4-4.

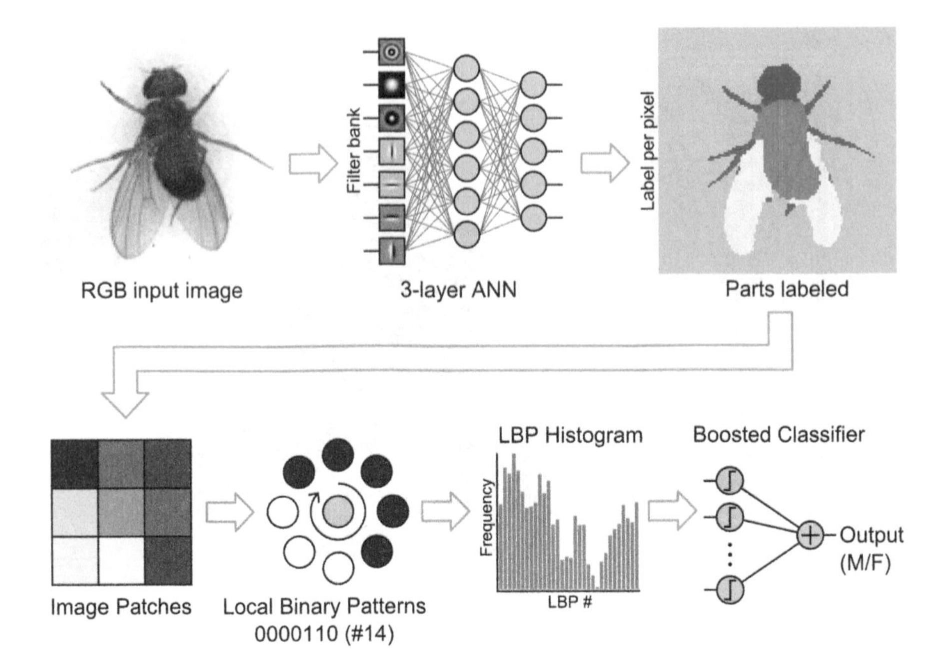

Figure 4-4. Sex classification algorithm

Close, But...

I took a step back to evaluate how well our approach worked. On the plus side, the prototype was relatively compact, not too expensive (around $1,200 in parts), and sorted flies fairly well. After many iterations on the code, Matt and I were able to achieve 97% accuracy sorting flies by sex.

Unfortunately, there were some real limitations. The design required users not only to empty a vial of flies onto the pad, but also to spread them out so that each was a few millimeters away from its neighbors. This step (which prevented two flies from getting sucked up at once) took a minute or two for a vial of 50 flies.

And there was the issue of accuracy. While 97% accuracy doesn't sound too shabby, that translates to an average of three mistakes per hundred flies. Sorting by sex is commonly followed by making a cross for breeding, and with a spurious male in a vial of females, that resultant cross wouldn't be worth much. To have a viable product, we needed closer to 99.9%, or better.

We considered our options to improve accuracy, and ultimately decided all the changes that might work involved trade-offs that would render the final product either too expensive, slow, or cumbersome to use. We'd hit a dead end.

A New Strategy

When we reconsidered the decisions leading to this fly sorting cul-de-sac, there was one that begged to be revisited: should we image awake or anesthetized flies? Flies under anesthesia don't move around, of course, but are challenging to work with. Their legs get stuck together like Velcro, and the flies are difficult to place in a consistent orientation. As we considered ways to image nonanesthetized flies, we learned that someone else—a group at ETH Zurich—had successfully built a sorter! Reading the journal article about their system, dubbed the FlyCatwalk, was simultaneously exciting and gut-wrenching (Medici et al, 2015 (*http://bit.ly/medici-flycatwalk*)). Although the article validated the purpose and feasibility of our project, someone had beaten us to the punch.

On the third or fourth read through the paper describing their very clever system—they imaged flies walking in the narrow gap between two clear plastic sheets—a few things became clear. First, using nonanesthetized flies was the way to go. It made the computer vision and machine learning code a lot simpler. And second, the FlyCatwalk had much room for improvement. Their system is quite large, taking up considerable space on a bench. Additionally, the throughput of the FlyCatwalk is quite low, about one fly every 40 seconds, and some flies never even make it out of the vials to be imaged.

Prompted by their success, and considering the limitations of their system, Matt and I discussed different ways we could usher individual, awake flies into an imaging chamber like the FlyCatwalk. We hit on one that sounded promising, essentially a fly "valve" made from soft foam rollers, and thus began another round of designing and building.

More than 50 prototypes later, we'd created a small device to automatically dispense individual flies, awake, one every few seconds. There were, of course, many details to work out—how to avoid dispensing more than one fly at a time, how to avoid damaging flies, how to make it work quickly, etc.—but ultimately the machine hewed very close to our valve idea, using two motorized foam rollers to create a barrier with flow control for flies (see Figure 4-5).

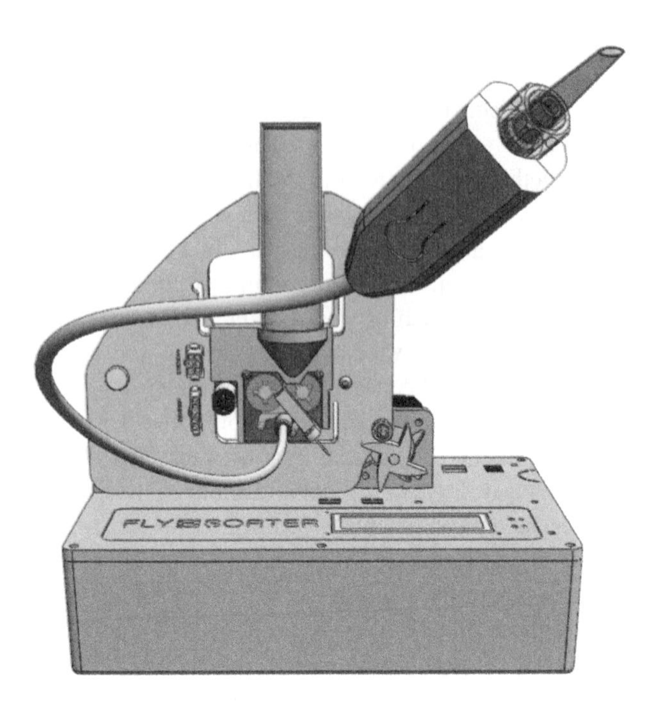

Figure 4-5. Automatic fly dispenser

The dispenser is a big step toward a high throughput classification system. As an added bonus, we've connected with researchers who study fly behavior and learned that some are eager to use the dispenser on its own, without any imaging at all. Loading up their experimental assays one fly at a time can take a long time, and this device not only speeds up the process, but it also avoids any potential side effects from anesthesia or human chemistry (by way of the mouth pipette). In start-up terms, we've found our minimum viable product, and by the time this article is published, I expect we'll have made our first sale.

What's Next?

We're glad to have built relationships with several high-profile labs around the country; their feedback has been and will continue to be extremely valuable, and the possibilities with collaboration are very exciting. It's going to be a busy six months at FlySorter. In addition to bringing the dispenser to market, we plan to prototype the next-generation classification system, and we have a few other projects in the works as well.

While I never imagined I would be wrangling fruit flies this way, I'm glad to have found this overlap between engineering and science. It's satisfying and rewarding to launch the results of two years of work publicly, and I'm excited to take on the challenges ahead.

Dave Zucker *is a jack-of-all-trades engineer in the Seattle area. He worked on projects large and small, from computer mice to robot bartenders, before founding FlySorter (http://flysorter.com/). You can find more information on the website and on Twitter (@FlySorter). For more behind-the-scenes stories, visit the blog Sex My Flies! (http://sexmyflies.com/)*

Standards for Protocols: The Quickest Way to Reproducibility

Ben Miles and Tali Herzka

Anyone who has worked in a lab is familiar with protocols that simply do not work as they should. Protocols are often subjective, written for oneself to use again in the future in the same lab. However, problems inevitably arise for the next scientist, whether in the same or a different lab, attempting to replicate the experiment that's missing some critical details about how exactly to do that tricky incubation step. Looking at the literature today, there are many examples of methods sections that are overly edited and ambiguous. The loss of precise descriptions of methods most likely occurs out of necessity to edit articles to an appropriate length for the publication; however, precision and reproducibility should not be sacrificed for better word counts. Table 5-1 illustrates a number of typical examples, taken from the literature, of vague descriptions of processes that are open to interpretation. This lack of precision in documentation hasn't gone unnoticed in the wider scientific community and is recognized as one of the causes of the widespread irreproducibility of scientific studies that is often discussed in the media. Thankfully, a solution to this problem is emerging in the form of standardized protocols.

Standard specifications can be used for describing protocols to eliminate their subjectivity and, as a consequence, variability. Picture a world where every protocol written uses the same format and terminology, where the protocol documents every step unambiguously. Protocols can be reworked, revised, and remixed with other protocols that adhere to the same standard to create completely new processes or refinements of older ones.

Right now, people are trying to answer difficult questions about how we as a scientific community can codify our complex and diverse documentation of how to pipette, incubate, take spectrophotometric measurements, and generally perform operations in the lab. Especially now, in the era of roboticized scientific platforms (see BioCoder #8 (*http://bit.ly/biocoder-8*) for more on robotic platforms), researchers have found themselves craving standardized instructions that robots can interpret. Standards like Autoprotocol and projects like Antha and Wet Lab Accelerator have begun to fill this gap for robotics, and it is increasingly becoming possible to extend the usage of standards to the documentation of manual lab work. In this article, we will cover which protocol standards exist, how we think they should be governed, and generally envision a world free from ambiguous prosaic protocols, where well-defined, repeatable, and shareable protocols enable unprecedented scientific developments and new levels of collaboration.

Table 5-1. Examples of ambiguous methods instructions from the literature alongside an explanation of their ambiguity and problems that can arise from this

Ambiguous Instructions from the Literature	**Problems**
Samples were run on 10% Bis-Tris gels (Invitrogen) and stained with Coomassie blue stain (Norville, 2011 (*http://bit.ly/norville-purification*)).	SDS-PAGE is fairly routine, and people know how they usually do it; however, for reproducibility it should be known how the authors conducted the SDS-PAGE. In this example, no indication on volume of sample being analyzed is given. Furthermore, no indication of the stain concentration, staining time, or volume is given. All of these will cause some degree of variability for anyone reproducing this analysis.
Strains were grown in YPD supplemented with 10mM Pi and 200μg/μL G418 at 30oC for 26 hours and then diluted 30-fold in synthetic complete medium with 10mM Pi and **allowed to re-enter log phase** (Rajkumar, 2012 (*http://bit.ly/rajkumar-eukaryotic*)).	Re-enter log phase is open to interpretation. When replicating this method, one researcher's interpretation of log phase may be at a greater microbial concentration than another's interpretation. Standards would define this instruction to incubate the culture for a fixed period of time post dilution.

What Is a Standardized Protocol?

Let's show you a standardized protocol in Table 5-2.

Table 5-2. The top panel shows the natural language prose description of a protocol, and the bottom panel shows a python script written to generate Autoprotocol JSON. The Autoprotocol JSON for this protocol produced by the python script can be viewed on GitHub (https://gist.github.com/bmiles/a9a4ecf9f32448e2a348).

1. Dispense lb-broth-noAB to 1 column(s) of growth_plate.
2. Transfer 2.0 microliters from DH5a/0 to growth_plate/0.
3. Transfer 2.0 microliters from DH5a/0 to growth_plate/12.
4. Transfer 2.0 microliters from DH5a/0 to growth_plate/24.
5. Transfer 2.0 microliters from DH5a/0 to growth_plate/36.
6. Cover growth_plate with a low_evaporation lid.
7. Incubate growth_plate at 37 degrees celsius for 12 hours.
8. Measure absorbance at 600 nanometers for wells A1, B1, C1, D1, E1, F1, G1, H1 of plate growth_plate.

```
 p = Protocol()
 # Set up the containers being referenced
 ctrl_bacteria =
p.ref("DH5a",id="3209xifd029",
cont_type="micro-1.5", discard=True)
growth_plate =p.ref("growth_plate",id="3208kfhf394",
cont_type="96-flat", storage="cold_4")

# Dispense, transfer, cover the plate, incubate, measure OD600
p.dispense(growth_plate, "lbbroth-noAB", [{"column":"0", "volume":
"150:microliter"}])
p.transfer(ctrl_bacteria.well(0), growth_plate.wells_from(0,4,columnwise=True),
"2:microliter")
p.cover (growth_plate,lid="low_evaporation")
p.incubate(growth_plate,"warm_37", "12:hour", shaking=True)
p.absorbance(growth_plate, growth_plate.wells_from(0,8, columnwise=True),
"600:nanometer", "OD600")
```

A standardized protocol is a scientific protocol expressed as a data structure such that it adheres to a certain documented specification and its steps are as clearly defined and as unambiguous as possible. A protocol specification defines how a list of ordered instructions should be expressed, what parameters exist for each step, and what values are acceptable for each of those parameters. Following a protocol specification is not just about how the protocol is expressed; the format allows for ease of version control and for further applications such as the translation into natural language, display as a graphical interface, programmatic comparison, and, importantly, execution by robotic automation.

The benefits of protocol standardization as it relates to the scientific method are clear: the more easily understood and descriptive a protocol is, the less researcher bias is introduced and the more confident we can be in results obtained by two different entities executing the same set of instructions. Likewise, the more easily a protocol can be shared and translated among researchers, the more often and accurately those experiments can be reproduced.

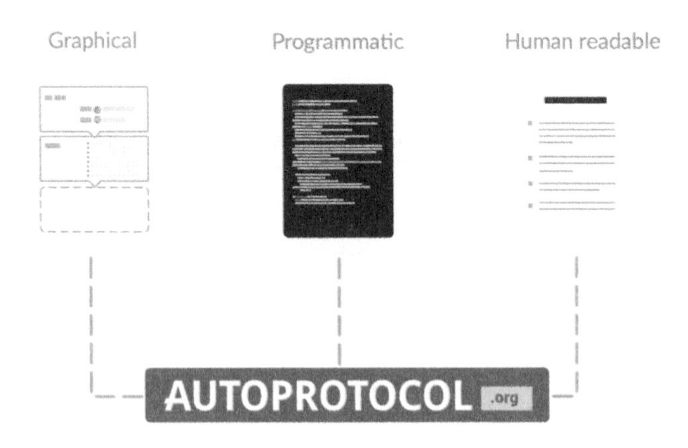

Figure 5-1. This schematic representation of the autoprotocol ecosystem demonstrates the value of having a standardized format for defining protocols. With Autoprotocol, computer programs or graphical user interfaces can be used to create protocols in the Autoprotocol format—in turn, autoprotocol-formatted protocols can be transformed into human-readable lab instructions, submitted directly to a robotic experimentation platform, or be visualized as a graphic protocol. Autoprotocol acts as a single source of truth for a protocol, which can then be interpreted or compiled into an output.

Standards

Adhering to standards, whether designing mechanical parts or formatting references in a thesis, is often complicated since you must decide upon which standard to use. Decisions on which standard to choose often take into account compatibility with existing components, the expected lifetime of a standard, and, more often than not, the ease of adhering to that standard.

First and most importantly, any standard that is adopted by the scientific community must be open. That is to say that the development of a standard should not occur in isolation from its users. The standard should be transparent and accessible to everyone—this is in line with some of the core principles of the scientific community contributing back to our shared knowledge.

There are a number of open approaches to standardized protocols already in existence. For example, Autoprotocol (*http://autoprotocol.org/*) provides an open standard that specifies a formal, JSON (*http://json.org/*)-based (a form of standard data structure itself) data structure for describing a protocol. Another type of standard is Antha (*https://www.antha-lang.org/*), which is a programming language that provides a high-level abstraction to experimental design for the biosciences. By Antha's nature of being a programming language, it is standardized, and upon execution of the code it generates a standard machine-readable protocol. Both Antha and Autoprotocol are in their infancy and will grow with their adoption. Both being open should ensure their improvement and development over time.

The existence of these and other standards hints at the growing need for scientists to communicate in a common language. Although expressing a protocol in a computer-readable standard may seem counterintuitive in terms of facilitating human collaboration, representing a protocol as a data standard at its core allows for many different kinds of abstraction on top of it while retaining the base description of instructions. In the case of Autoprotocol, its flexibility as a standard has already been demonstrated. Individuals can hand write protocols using the Autoprotocol structure itself, generate Autoprotocol using Python with autoprotocol-python (*http://bit.ly/autoprotocol-py*), or create Autoprotocol-adherent protocols graphically using Autodesk Inc's We (*http://bit.ly/wetlab-accelerator*)t Lab Accelerator (*http://bit.ly/wetlab-accelerator*). A screenshot from the Wet Lab Accelerator is shown in Figure 5-2, where its simple drag-and-drop interface allows for the arrangement of steps into a visual protocol that can then be compiled into Autoprotocol for other uses.

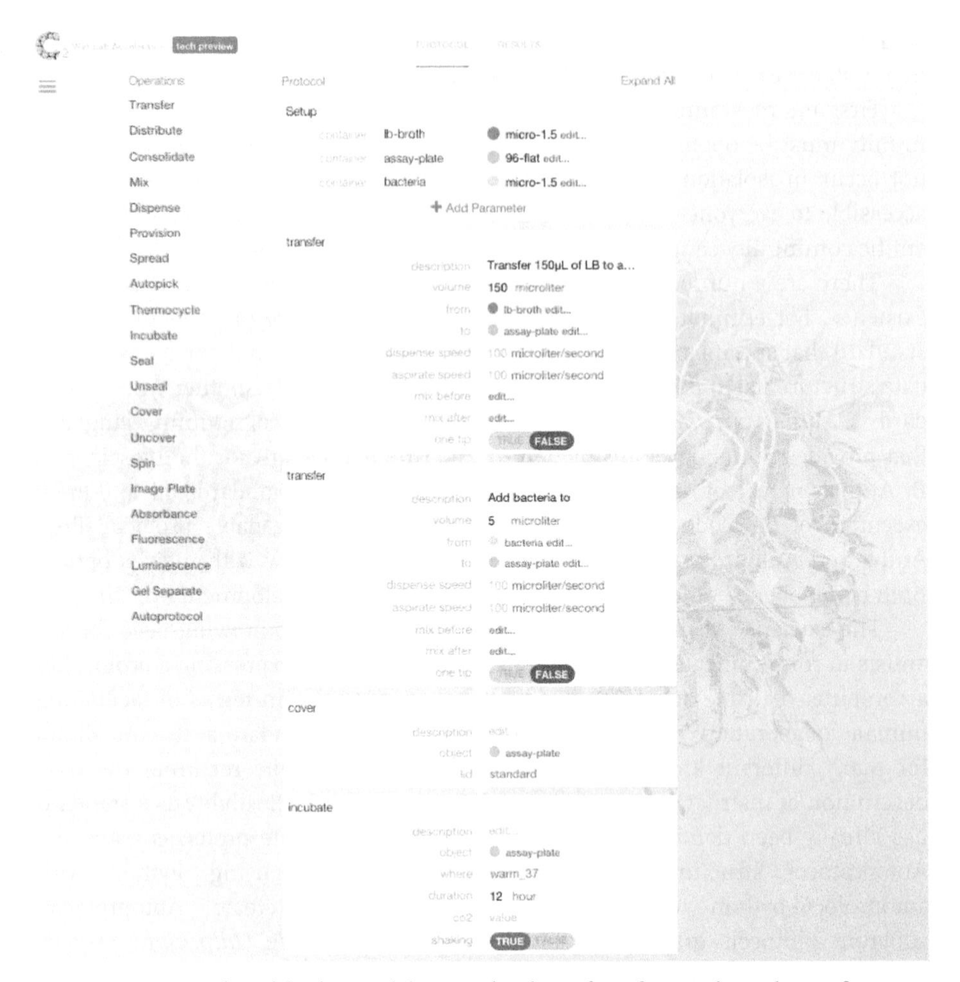

Figure 5-2. A screenshot of the drag-and-drop graphical interface of Wet Lab Accelerator from Autodesk Research. Users simply drag and drop different experimental instructions into a chain. Parameters, such as microplate wells, pipetting speeds, and volumes, can be set on each instruction card. Protocols created using Wet Lab Accelerator can be exported to a file as Autoprotocol JSON for other applications or be submitted straight to the Transcriptic robotic cloud science platform.

There are also forms of structured protocols that don't adhere to any specification. For instance, the website *protocols.io*, allows users to document and share protocols through their web platform. When creating a protocol on protocols.io the user must enter the protocol as a set of steps that fall within the data structure defined by protocols.io. This takes protocols toward a standardized format beyond just prose in a methods section. However, while the protocols are documented within a structure, the content of each protocol step does not itself adhere to a standard specification.

Beyond the implementation of protocol standards, there are some interesting discussions to have about how they should be governed. For Autoprotocol, there is a process in place for making modifications; however, the standard is too young for it to have a formal governing body. Autoprotocol Standard Changes (ASCs) are proposed to the community and discussed via a developer mailing list, similar to the way much larger open source projects operate. People who are not official curators must sign a Contributor's License Agreement (CLA) in order to have their change considered. For now, the standard is maintained by a small group of curators, but as the standard grows there will be a need for further subdivision, with committees responsible for duties such as maintaining the overall structure of the standard itself and determining the utility of proposed changes.

There is much to be learned by open protocol standards from other open source projects like the Apache Software Foundation, Openstack, Ubuntu, or Openstreetmaps in regards to governance. One can imagine that the process of incorporating changes to a protocol standard will work similarly to the way the rest of science works, with contributors coming to a consensus about how elements of the standard should be expressed and paradigms shifting as scientific techniques change and expand.

Open Science, Open Protocols

Open source protocols are protocols that are freely accessible to anyone and anyone has the right to modify them. These partially exist today, but lack of structure fails us as scientists. Methods are routinely described in open access journals like PeerJ (*https://peerj.com/*) or websites like www.openwetware.org and www.protocols.io. Of these methods for sharing protocols, protocols.io is the only one that has a well-defined process for the duplication and adaptation of shared protocols. The process used by protocols.io is analogous to that seen in the open source software community on GitHub (*https://github.com/*).

Currently, protocols are viewed as the byproduct of a research project, or the documentation of steps undertaken to reach a result. However, reuse of a protocol by other researchers is often an afterthought. Introducing protocol standards makes strides toward productizing protocols, and once protocols can be seen as a research product to aim toward, this will better align the incentives of researchers to create reusable protocols that are of higher value due to their reproducibility. Once protocol value and incentives become realigned, the creation and sharing of protocols will become increasingly used by researchers and funding bodies alike to gauge the impact of a project and its supporting grant. All of this will combine to make it more attractive to create and release new protocols for the scientific community to implement with confidence.

Collaborative Protocols

GitHub is a website that hosts computer programs that people have written and facilitates the sharing of code in a way where it can be copied and modified easily. GitHub also allows for the merging of modified code back into the original source as an improvement. One of the most noteworthy outcomes of the introduction of GitHub to the computer science world was its widespread inspiration of collaborative creation. Groups or individuals from around the world have been enabled to write and share code and improve upon existing software. This is the direction in which protocols will be heading. Currently, it is difficult to collaborate on protocols; one person's meaning of "mix-well" or "resuspend" could be completely different from somebody else's definition of the same action. However, in a world where everyone is speaking the same "language" of protocol, suddenly each step has a semantic meaning, and additions to the protocol sit harmoniously alongside the original steps.

A universal specification of protocol language means that people within the same lab or across institutions will be able to collaborate on the improvement and development of existing protocols. They will be able to look back into a protocol's history and see the exact moment that extra rinse step was added and by whom. These changes could be correlated with standard validation tests that may be run for a particular protocol to regularly gauge any improvement made to it.

There is a possible future where there are websites like GitHub hosting versions of standardized protocols that researchers can duplicate and change or submit improvements back to the original source of the protocol. A whole community of individuals collaboratively working on reproducible protocols to be used throughout the scientific community could flourish. Academic laboratories could publish their protocols to an online repository at the same time they publish a paper. The paper conducted with version 1 of the protocol, improved continuously to version 3.5 in the two years since the original paper. New work, standing on the "virtual" shoulders of giants prior.

Aside from global collaborative open science, there are more immediate benefits to more familiar collaborations such as interacademic and industrial research partnerships. Researchers are often geographically separated from their collaborators, and the sharing of research outputs is limited to sharing samples or data. The transfer of a traditional protocol is fraught with issues around interpretation of the text and performing the protocol in a different lab with dissimilar equipment and different environmental conditions. With standardized protocols and similarly standardized equipment, the reproducibility could be greatly increased, which would impact these academic industrial partnerships very positively.

Future Directions

We want to see a future where journals recognize and take partial responsibility for sharing irreproducible protocols. To this end, publishers should recognize the importance of standards to protocols to the point where when it comes to submitting the methods for a paper, the method must comply to a known standard.

There are, of course, challenges facing the adoption of standards for protocols. For instance, even if a protocol adheres to a standard structure, it is possible that the instructions in that protocol are incorrect. In this case, who is responsible for validating a protocol? It is the author's desire that existing methods publications shift their focus to sharing validation data and discussion while the protocol exists in a repository similar to GitHub or protocols.io.

Furthermore, it is often very difficult to get people to adopt a new way of doing things, and adding specifications to protocol documentation will add friction to the process of documenting methods. The way to address this is to provide immediate value for writing standardized protocols. This perceived value could be in rewarding research that outputs standardized protocols, or it could be more immediate incentives of facilitating easier collaboration with research partners or facilitating the easy execution of experiments on robotic research platforms.

Conclusion

Widespread irreproducibility of protocols is a systemic problem in the life sciences. Describing protocols is one of the places where scientists can take control of the reproducibility of an experiment. A number of standards are beginning to take shape out of necessity in the lab automation space; however, these standards if applied correctly to traditional wet lab work could drastically aid scientists in academia and industry alike. There exists the desire to share protocols among researchers as demonstrated by journals and other online tools; however, the scientific community lacks the necessary tools to express protocols properly while encompassing all of the essential details required for their successful execution. With the introduction of a standardized format, protocols can be shared and reproduced easily. The creation of new standards is in its infancy; the adoption and input of scientists is essential to the development of a genuinely useful tool that can reduce irreproducibility and improve collaboration.

Ben Miles *is a post-graduate student at Imperial College London in the Institute of Chemical Biology's Centre for Doctoral Training. His soon-to-be finished PhD covers topics such as bio-mineralization, extracellular matrix proteins, and automated flow chemistry. He blogs on technology, science, and startups at www.benmill.es.*

Tali Herzka *is one of the original contributors to Autoprotocol, an open standard for scientific protocols and automation. After recieving her B.Sc in biology from McGill University, she spent three years at Cold Spring Harbor Laboratory. Tali then joined Transcriptic Inc., where she is an application scientist working to create reproducible protocols using Autoprotocol.*

Vertical Farming and the Revival of the California Exurbs

Rebecca E. Skinner

Introduction

The cities of the near future will house more people, under more constrained land and water resources, than we have experienced thus far in the twenty-first century. Accommodation to these restrictions will involve infrastructure—for transportation, housing, and management of coastal waters—that is both large in scale and novel in concept. This article suggests that vertical farming is one such opportunity, suited to major world cities and beyond. It is already emerging as a popular option in world centers; we are concerned with its possible implementation in blighted exurban areas, for instance those of California, where it could help tremendously.

Vertical Farming: A New, Old Technology

Vertical farming, which is farming done in buildings constructed or retrofitted for this purpose,[1] is a venerable idea whose time has come. Urban farming is not new, having apparently taken place in Soviet Armenia in 1951. It was suggested as early as a 1909 article in *Life* magazine (see *Wiki* entry, June 3, 2015). More recently, urban farming has been suggested as a panacea to logistics costs of industrial farmed goods and a force to counter urban pollution. Its most famous current advocates are Columbia professor Dickson Despommier and Malaysian architect Ken Yeang, but vertical farming is so widespread that it is both an urban planning concept and a popular movement. This article is more concerned with its broader uses for large-scale farming than with its admitted potential aesthetic appeal.

1 Wikipedia cites the first formal definition by G.E. Bailey in 1915.

Vertical Farming Illustrated: the Vertical Harvest Project

The benefits—and challenges—of a vertical farming system are illustrated by discussing any one of a number of projects. We will use illustrative photos provided by the Association for Vertical Farming. As Figure 6-1 and Figure 6-2 (which show a VF vegetable garden) indicate, the central requirements are farming in a building—often a repurposed industrial one—rather than in the ground. The fact of an enclosed space immediately offers the opportunity of significant water conservation and near-elimination of pesticide and herbicide requirements. Plants may be watered conventionally or sprayed at the roots, a technique known as *aquaponics* that was developed in conjunction with NASA.[2]

Figure 6-1. An Association for Vertical Farming rooftop garden (photo courtesy of H. Gordon-Smith, agritecture.com)

2 Despommier p145, reference to aeroponic watering developed by Richard Stoner while working for NASA.

Figure 6-2. Vertical farming greens (photo courtesy of H. Gordon-Smith, agritecture.com)

Additionally, the enclosed space, lacking the full-sun exposure of an open field, demands the significant requirement of a light source. The light source may be at least partially natural. This is quite feasible in a repurposed industrial building, and remedies to the inevitable costs are currently under development by Dutch industrial conglomerate Phillips, which contributes significant R&D to LEDs specifically for vertical farming.

Once the lighting challenge—acknowledged as one of the most significant impediments—is overcome, the design possibilities are immense in terms of both labor-saving and space efficiency. Often stackable trays that may be moved into place by hand or on automatic mechanisms are used. This basic design has been patented as a "hydrostacker." This means that the operation is considerably, but not entirely, roboticized.

A synthetic environment offers an ideal opportunity to redress the sundry deficiencies in various places, which used to be far more attractive for farming. It is always too cold and dry for winter farming in Wyoming; there will always be inadequate water and a surfeit of heat in much of the Middle East; ambient plant parasites and insects kill crops everywhere; human pathogens threaten field workers. Nitrogen runoff from agricultural fertilizers poisons water by encouraging algae bloom. There is inadequate sunlight during much of the year in the northern cities of North America. Artificial light offers optimal summer-like lighting any time of the year, regardless of latitude.

There are dozens of such projects. Other large efforts are underway at Growing Power in Chicago; London's Thanet Earth, which provides more than ten percent of the salad greens for England; Bright AgroTech in Laramie Wyoming; 10 Mile Farms in Las Vegas and Tokyo (where the Pasona office building shares the farm with a temp staffing agency); the Vertical Harvest project in Jackson, Wyoming; the Farmed Here commercial produce facility near Chicago; a 2011 project in Suwon, South Korea; and Plantagon, an elaborate greenhouse in a geodesic dome, construction of which is underway in Lingkoping, Sweden.

Other Reasons to Do Vertical Farming

Practically every journalistic illustration of VF is oriented toward green vegetables, which are impossible to grow in winter. In some ways this adaptation of urban areas to a very different land use is a more technically sophisticated version of existing community gardening and urban farming movement. Recent attention to vertical farming aside, agriculture has been in the process of remaking itself in distinct independent ways for decades. These new approaches include hydroponics, which was developed in the USA largely for growing marijuana indoors and in desert areas of Israel for water-efficient farming; indoor farming for year-round crops in cold places; the widespread community gardening movement; the tilapia farms of Growing Power in Milwaukee; worldwide urban homesteading; and truck gardening (often high-end and organic), which is beloved by the locavore and slow food movements.

At least as interesting, is the prospect of vertical industrialized agriculture that can take on industrial needs as well. The raising of plants need not be attached solely to conventional food production. Medical marijuana certainly should be grown under more rigorous settings than the current growing arrangements in the Emerald Triangle, much of which is a patchwork of illegal fields, often in national forests using stolen water. Vertical farms could provide a great environment for THC and marijuana production under highly controlled conditions. Even repurposed concrete buildings could be excellent sites for various projects such as mycology, 3D printed foods and animal products, algae biofuel, and possibly even a vegan "hamburger" that tastes good. Tobacco raised under controlled conditions offers the substratum for flu and Ebola vaccines.

The Central Valley and Delta: Regions in Search of a New Regional Economy

The showpiece projects of vertical farming thus far have been largely located in major urban areas. However, the practice offers a great deal to exurbs as well. Farming industrialized through vertical farming and scientifically informed design could not meet up with a more deserving set of circumstances than those of the region just to the east of Silicon Valley. San Jose and Stockton are 86 miles apart, and despite their proximity, the contrast between the two is stark. Stockton, population 696,000, located along the highly polluted San Joaquin River, lingers at the bottom of many measures of wellness, not only in California, but on a national level. These include obesity, percentage of population with higher education, municipal fiscal health, employment figures, drug production and use, and environmental health. After going bankrupt in 2012, the city only exited bankruptcy status at the end of February 2015. Its largest employers are San Joaquin County and the Diamond Food processing company. The fiscal structure of the situation—property tax receipts collapsing because of underwater homes, and Muni bonds issued just to refinance other Muni bonds, rather than to fix infrastructure (*WSJ*, May '15)—suggests that these are deeply rooted problems. Yet these problems are not unique to Stockton, but are found in many comparable California cities, especially those of the adjacent region.[3]

While farming appears attractive to day trippers, in the Central Valley and Delta it is beset by serious problems, many of them beyond the scope of local policy remedies. Bluntly stated, the economic and environmental health of the region is catastrophic. There's not enough water, and what dwindling supplies there are, are hotly contested. The area's farming is also eroded by land subsidence and environmental pollution from pesticide application and from methamphetamine manufacture. The increasingly fractious and ugly issue of water rights continues to grate, as junior water-rights holders are entirely cut out of a smaller supply by senior water-rights holders (*NYT* June 6, 2015 "Sipping California Dry"). This leads to land subsidence and the threat of salt-water flooding back into the Sacra-

3 Vallejo, which was bankrupt between 2008 and 2011, has 115,000 people, Modesto half a million, Bakersfield 840,000, Fresno almost a million, and Visalia-Porterville 450,000 ("Vallejo" and "Central Valley," Wikipedia, accessed June 3, 2015).

mento–San Joaquin Delta. Many farmers continue to pump out groundwater, both their own and water from their neighbors' land. These rights were further curtailed by the state's water rights authorities in mid-June (*NYT* June 13, 2015). River-water diversion for farming is a political hot potato that will continue to nag the Central Valley.

Health concerns from both farming and from the drought's impact on soil in the Valley and environs are drastic. The usage of pesticides on crops inevitably leads to ingestion and almost certainly health problems such as asthma and Parkinson's disease for farm workers (as documented by Cal-EPA and other studies). A side effect of the Valley's historic poverty is the production of methamphetamine. The region is acknowledged to be the US's most active location for this drug, which is dangerous in many ways and underlines the underusage of existing land (Winter, 2011). Last, but by no means least, is Cocci fever (Coccidioimycosis), a dust-borne human pulmonary illness that is increasing in incidence in the Central Valley and Kern County, and the American Southwest generally (CDC *Cocci Fact Sheet*, 2011).

Vertical farming offers further advantages that complement California's specific and dire needs well. Simply by reason of being enclosed and using water in a highly measured and judicious manner, vertical farming could reduce water usage tremendously. This method offers crops year-round, regardless of heat waves or air pollution, or Tule fog and drought (in California).

Vertical farming, undertaken in large scale in the cities of the Central Valley and Delta, could help to revive local economies. The labor structure associated with vertical farming would not resemble the farms of the Central Valley historically. The latter economy is notorious for low wages and labor violations, but it certainly employed huge numbers of people. This is not necessarily so with a highly roboticized and automated factory design. Yet its construction, administrative offices, distribution warehouse hubs, and transit activities would still employ thousands of people and offer extremely high-volume production of goods, as well as informational externalities associated with the management of farms. The anemic cities in the Central Valley and Delta would be ideally situated to this new purpose. They have cheap urban land and buildings, many people in need of work, and are relatively near both the greater Bay Area and SoCal.

Conclusion: Reviving Vallejo (and Stockton and Modesto and Tracy and...) with Vertical Farming

Like more distant regions of the US that the tech world rarely glances at, the problems of the outer exurbs of the Coastal Bay region appear intractable. Yet tech offers a brighter alternative future, one of many points where economic desperation meets technological opportunity (or a great surfeit of labor meets much-needed capital). Large-scale vertical farming and the distribution logistics built around it could help revive these cities. After all, it's only 86 miles from Stockton to San Jose.

References

- Cockrall-King, Jennifer. *Food and the City: Urban Agriculture and the New Food Revolution.* Amherst, NY: Prometheus Books. 2012.
- Despommier, Dickson. *The Vertical Farm: Feeding the World in the 21st Century.* New York: St. Martin's Press. 1999.
- D'Ordorico, Paolo. "Feeding humanity through global food trade." *Earth's Future*, 2014; DOI: 10.1002/2014EF000250.
- Iozzio, Corinne. "A Farm on Every Street Corner." *Fast Company.* April 2015: p68.
- Fabian Kretschmer and Malte E. Kollenberg. "Vertical Farming; Can Urban Agriculture Feed a Hungry World?" *Spiegel Online.* July 22, 2011. *http://www.cityfarmer.info/2011/07/23/south-korean-city-of-suwon-has-a-vertical-farm/.*
- Marks, Paul. "Vertical Farms Sprouting all over the World." *New Scientist.* 16 January 2014.
- Phillips LE, website.
- The Plant Chicago. http://www.plantchicago.com. Also Cockrall-King p274–282.
- Stech, Katy. "Stockton, Calif. to exist Bankruptcy Protection on Wednesday." *Wall Street Journal*, February 24, 2015.
- *The Economist.* "Vertical Farming: Does it really stack up?" Q4 2010. December 9, 2010 Print Edition.
- Vertical Harvest, website and news materials.
- Winter, Ray. "New factories in the fields." *Boom: A Journal of California.* Winter 2011.

- Zumkehr, Andrew and J. Elliott Campbell. "The potential for local croplands to meet U.S. food demand." 2015. *Frontiers in Ecology and the Environment* 13: 244–248.

***Rebecca E. Skinner** is the founder of Zoetic Data, a personal environmental applications start-up based in San Francisco. Her books, Building the Second Mind and The Dante Machine, are available on Amazon and eScholarship. A former Stanford University research associate and lecturer, she holds a PhD in City Planning from U.C. Berkeley.*

Putting the "Tech" in Biotech

Edward Perello

We were promised better software tools. They would assist the rational engineering of biological systems. While there has been great progress, there remains tremendous scope for us to do better. The tech community (by which I mean those involved in software development) has grown rapidly and changed the world in just a few short decades. If the attention of that community can be directed and nurtured toward building new software tools for synthetic biologists, then we are likely to see similarlyrapid progress in the development of the tools our field requires. Unfortunately, the tech community does not "speak biology"—few are even aware of the developments happening in biotech. If we are to build the bio-coding community to its fullest potential, then we must engage "tech" directly, and involve as many traditional software developers as possible. I propose a method to add more strings to the bow of synbio by directly engaging the tech community, training them in biology, and getting them to solve real problems that biologists face.

The State of Biology Software Today

The UK's Synthetic Biology Leadership Council projected that by 2015 our field would be benefiting from a host of academic labs and companies developing and releasing tools to facilitate our work (Figure 7-1). In theory, the software tools these groups built would bring bioCAD and bioCAM technologies to the forefront of all biological engineering, and, in turn, these would let us move faster than human speed both inside and outside of the lab.

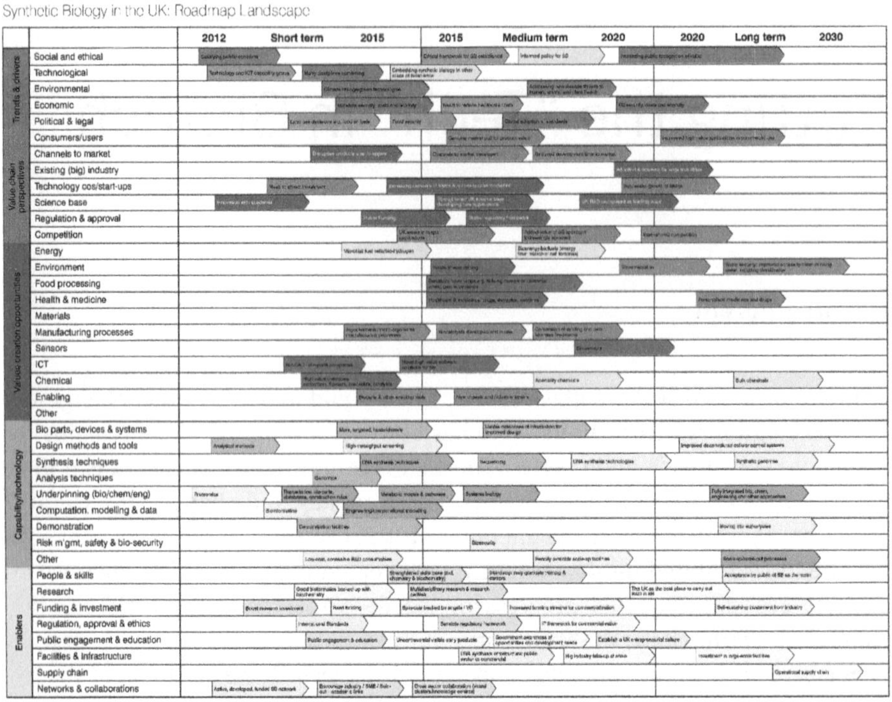

Synthetic Biology in the UK: Roadmap Landscape

		2012 Short term 2015	2015 Medium term 2020	2020 Long term 2030
Trends & drivers	Social and ethical		Informed policy for SB	
	Technological	Technology and ICT capability grows; Many disciplines combining	Embedding synthetic biology in other areas of innovation	
	Environmental		Addressing new disease threats to humans, animals and plant health	
	Economic		Need to reduce healthcare costs	
	Political & legal	Food security	Global adoption of standards	
	Consumers/users			Improved high value applications in commercial use
Value chain perspectives	Channels to market	Disruptive products start to appear	Chosen route to market developed; Reduced development time to market	
	Existing (big) industry			
	Technology cos/start-ups	Needs to attract investment		
	Science base	Innovation with academia	UK R&D recognised as leading world	
	Regulation & approval	Public funding	Stable regulatory framework	
	Competition			International competition
Value creation opportunities	Energy	Microbial fuel cells/biohydrogen	Bioenergy/biofuels	
	Environment			
	Food processing			
	Health & medicine		Healthcare & medicines: drugs, therapies, vaccines	Personalised medicines and drugs
	Materials			
	Manufacturing processes			
	Sensors		Biosensors	
	ICT			
	Chemical		Speciality chemicals	Bulk chemicals
	Enabling		New chassis and industrial strains	
	Other			
Capability/technology	Bio parts, devices & systems		Usable databases of information for improved design	
	Design methods and tools	Analytical methods; High-throughput screening		Improved decentralised cellular control systems
	Synthesis techniques	DNA synthesis techniques	Sequencing; DNA synthesis technologies	Synthetic genomes
	Analysis techniques	Genomics		
	Underpinning (bio/chem/eng)	Proteomics; Reproducible bio-parts, databases, construction rules; Metabolic models & pathways	Systems biology	Fully integrated bio, chem, engineering and other approaches
	Computation, modelling & data	Bioinformatics; Engineering/computational modelling		
	Demonstration	Demonstration facilities		
	Risk m'gmt, safety & bio-security		Biosecurity	
	Other	Low-cost, accessible R&D consumables	Readily available scale-up facilities	
Enablers	People & skills	Strengthened skills base (incl. chemistry & biochemistry)		Acceptance by public of SB as the norm
	Research	Good bioinformatics backed up with biochemistry	Multidisciplinary research & research centres; The UK as the best place to carry out R&D in SB	
	Funding & investment	Boost research investment; Seed funding	Spin-outs backed by angels / VC; Increased funding streams for commercialisation	Self-sustaining investment from industry
	Regulation, approval & ethics	International Standards	Sensible regulatory framework; IP framework for commercial value	
	Public engagement & education	Public engagement & education	Uncontroversial visible early products; Government awareness of opportunities and development needs; Establish a UK entrepreneurial culture	
	Facilities & infrastructure		Big industry take-up of areas	Investment in large-scale facilities
	Supply chain			Operational supply chain
	Networks & collaborations	Active, developed, funded SB network; Encourage industry / SME / Spin-out - academic links	Cross sector collaboration	

Figure 7-1. The UK Synbio Roadmap Landscape—pay close attention to the ICT section. Source: SynBio Roadmap Cordination Group, Research Council UK, 2012

So far, there have been some successes, and we see more appear every year. Transcriptic, Synthace, Desktop Genetics, Emerald, Riffyn, Benchling, Zymergen, Ginkgo Bioworks, Synbiota, Genome Compiler, Teselagen, Clotho, and the SBOL format—all of these are making concerted efforts to reduce the complexity of synbio and improve different aspects of the design-build-test cycle. The output of the companies, as well as every group presenting work at the IWBDA conference, demonstrates that great leaps forward *are* being made in the development of better tools—Not only in fundamental bioCAD/CAM algorithms and their operations, but also in their applications within whole ecosystems, perhaps most interestingly promised by Antha-lang and other high-level programming languages for biology.

However, the emerging strain of "biocoders" that can be found in these companies and conferences remains an incredibly rare breed. So, too, are the tools they build. One should keep in mind that the vast majority of biosoftware tools that exist out in the wild are either small-scope bioinformatics applications that come in the form of scripts attached to rudimentary user interfaces, or are alternatively sprawling, complex beasts developed in the early days of the web by biologists with little software development experience.

I would argue that while we have made progress, and that the UK Synthetic Biology Leadership council could *potentially* be satisfied by what we see today, there remains a dearth in useful, intuitive software tools that work *for* the biologist. Instead, we see a saturation of poorly thought-out generalist bio-tools that place functionality over form and in many cases impede a scientist's work rather than facilitate it. For the individual scientist, this means repeated work, wasted time, intense frustration, and slower research. More generally for the life science domain, this leaves us trapped in step-wise experimentation and optimization cycles alongside slow progress at both the academic and commercial level.

Too long have we wasted our time using frustratingly complex genome browsers and online forms that take 30 minutes to calculate a score (Figure 7-2 and Figure 7-3).

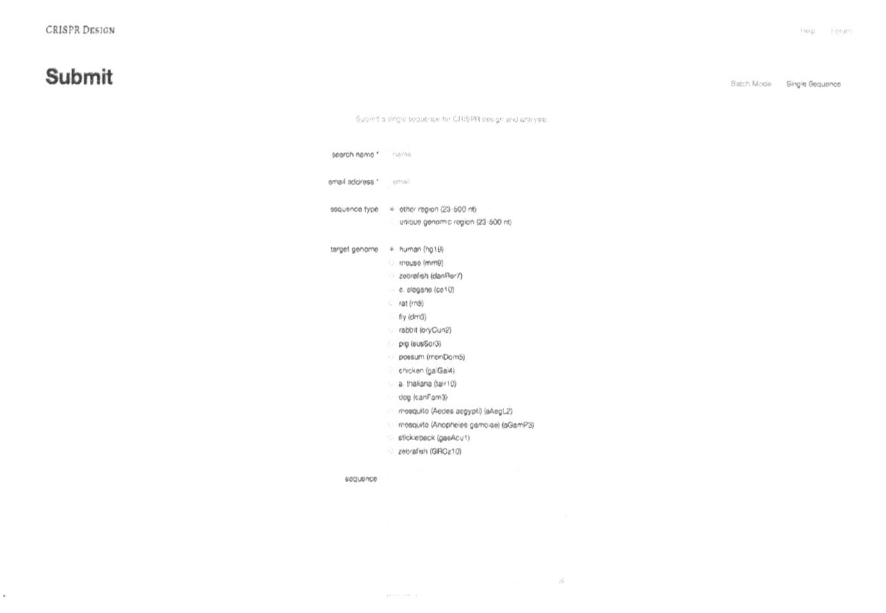

Figure 7-2. The MIT CRISPR tool, which is one of the most important tools in genome editing today. It houses powerful algorithms that find guide RNAs and score them based on the number of off-target hits each guide would hit within the desired genome. Nonetheless, it is slow and unwieldy—you can only input 250 base pairs per run, and it can take up to 30 minutes to complete each run, a classic example of user experiences ripe for improvement. Source: crispr.mit.edu —see Hsu et al., Nature Biotechnology 2013

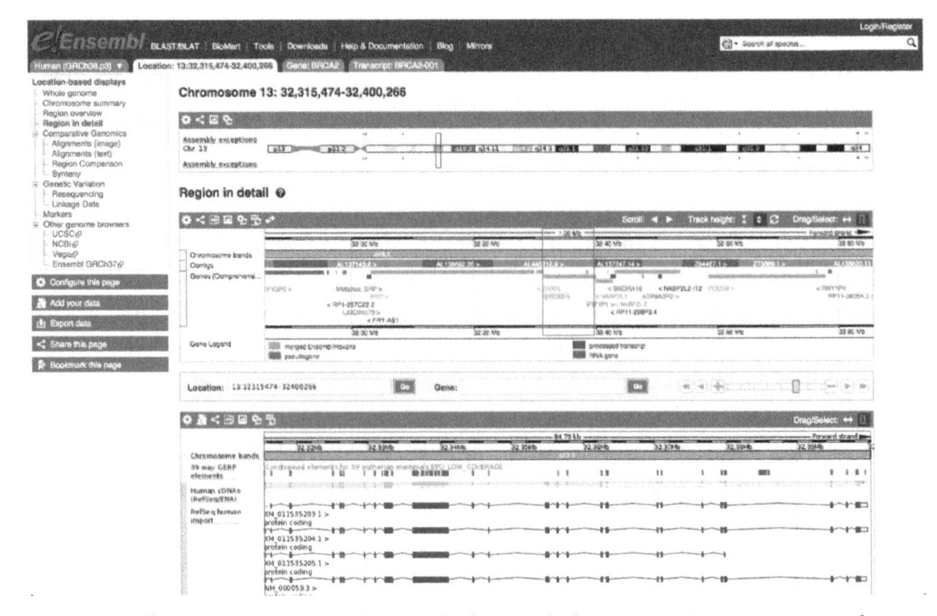

Figure 7-3. The ENSEMBL genome browser, looking at the human BRCA2 gene. A screenshot alone cannot describe the complexity of its operation. Instead, I encourage you to use it at ensembl.org. Source: Wellcome Trust Sanger Institute, European Bioinformatics Group, see Hubbard et al. (2002)

The successes I mentioned previously tease us of with a forthcoming renaissance in biology software, but a key problem has yet to be addressed: the small number of biocoders available to design and build the tools themselves. We need more software developers speaking the language of biology, and to achieve the future we look to create, I believe that we must cross-pollinate the synbio and tech fields as soon as possible.

Biologists Struggle with Existing Software Tools

The majority of biotech software tools are developed by academic biologists with little or no professional development experience. They are motivated by, and recognized for, the powerful algorithms and predictive models (great back-end power) they develop, rather than the user experience afforded to the user. As such, little attention is paid to design thinking, workflow optimization, or even creating tools that are "fun" to use (in other words, they often have poor front-end design). Such tools are not only a chore to use, but they are also poorly supported in the long term, becoming redundant as new browsers or operating systems are released.

The problem is ubiquitous. I have asked dozens of molecular biologists a simple question to which the answer is usually the same:

Q. *Can you think of a software tool in your lab that you use often, that really annoys you, but could be improved in a tangible way very easily? Pick something that you would describe as the bane of your existence.*

A. "*Yes, I have found that [I have a lengthy series of complaints to make about genome browser X and workflow system Y], and we have just had to get on with it and hope they fix it in the future, but no one has.*"

Occasionally, I also encounter biologists who think the software is well-designed but that the problem is their own because "their computer hates them."

The Opportunity and Challenge of Putting the Tech in Biotech

Software developers and user experience (UX) professionals have built technologies that have sped up our ability to work, communicate, and play in almost any situation imaginable. Bringing these professionals into the world of biology will yield similar enhancements in the synthetic biology field.

At this point, you might be thinking that we already have these multidisciplinary people, and that they are called *bioinformaticians* and *computational biologists*. These professionals bridge some aspects of biology and software, but I would point out that they are typically biologists who have picked up computer science—as well as many bad habits.

On the other hand, computer scientists and software professionals are trained in very different ways, and I have found that they bring fresh perspectives—they have a profound ability to think computationally, break problems apart, and simplify complex workflows in a unique way. When applied to the challenges of synbio, the expertise of a computer scientist could substantially improve biology researchers' ability to get work done. Indeed, synbio is so exciting precisely because it includes so many non-biologists who bring these new perspectives.

Nonetheless, there are challenges to involving them. A key barrier to a developer's participation in the field is the requirement to have a good understanding of biology itself. In my work to build Desktop Genetics, a company operating at the intersection of CRISPR gene editing and software development, we have found that developers, while eager to work in the synthetic biology field (it is very cool!), have trouble understanding the basics—let alone the complexities. Looking outside our company, most developers lack the requisite knowledge to rapidly translate a biologists' requirements into working code, and this can take months or years to develop.

That is not to say that it is impossible. In fact, we have put tremendous efforts into training our developers in biology, and we are incredibly excited to be building a team of developers who can hold their own in scientific conversations with post-docs about the specifications of a new feature, or even the scientist's reasoning for the importance of that feature. A problem we face is that such training takes an incredible amount of time because the learning curve is so steep. For us, this is our bread and butter—we are prepared to do this. But for academic labs and small biology companies, it is normally economically unfeasible to bring in software professionals for one-off projects, as the work to train them would cost more than the value of the work product itself.

There is also a school of thought that developers do not need to learn biology at all—rather they should be instructed and managed by a biologist, or someone who can translate biology to software easily. This method is used by larger companies and labs, but it stifles professional development and the acquisition of biology skills by those developers, which is precisely what we have seen to be important in generating multidisciplinary individuals for a multidisciplinary field. Further, this approach anchors the multidisciplinary individuals to managing a development team instead of being out in the field and identifying problems to solve.

Without doubt, developers who are trained in biology have become tremendously useful personnel within Desktop Genetics and other companies in this sector. These biocoders can move through the problem space of synthetic biology software very fluidly, and they can interact directly with a biologist (no need for "translation"); understand the context of the technical problem; empathize with the user's issue; work quickly to design a solution; test the proposed solutions; and code the solution itself. These biocoders can essentially conduct the software design-build-test cycle far quicker and more cost-effectively than a team where expertise is siloed away to the point where there are multiple points of failure due to communication and translation issues.

The Power of a Biocoding Revoloution

An ideal software tool condenses the knowledge and workprocesses of experts into a black box that can be used by a nonexpert. The end user is able to repeatedly call upon that expertise at will, anywhere in the world.

It's simple mathematics—more biocoders can work directly with more biologists and translate their expertise into tools that can be used by other biologists. An abundance of tools (that can communicate with one another and are easy to use) would facilitate reproducible scientific investigations and the development of powerful new biotechnologies. As more biocoders are trained, not only will we see a new wave of tools that can speed up work in the lab, we will also create a new generation of biologists who can program computers and life, work with start-ups, and start new companies and laboratories of their own.

The power of the biocoder is already recognized. They are often snapped up by major companies, resulting in a brain drain from the academic labs and the innovative start-ups that need them. Indeed, great interest has been shown in biotech by tech companies like Google and Autodesk, and it is clear that tech's interest in biotech will only grow over the next decade. Lets face it—biology is really just a new programming language, and frankly, these companies (and the people that make them up) already have a great framework to understand it.

More importantly, the tech community is hungry to get involved in the next digital revolution. As synthetic biologists, we already know what that revoloution will look like. But the tech community is only just starting to sink their teeth into it.

A Plan to Put Tech in Biotech

Training developers in biology and expecting them to hold on to that information in the long term is unlikely to yield the results we are after. Developers learn new programming languages best while practicing them on the job, and in my experience, small focused projects will teach trainee biocoders enough biology to communicate with a scientist, and it will inspire them to learn more biology on their own time in the future.

It is unnecessary and perhaps even counterproductive for trainee biocoders to do a Bachelors in biological engineering before letting them loose as independent consultants. Instead, I propose executing a two-step plan: 1) teach developers "a bare minimum" to understand what a biologist is talking about and 2) give them the resources to take part in biocoding projects and learn about real-world problems that they could solve with their skills.

Thanks to the support of the SynBio LEAP project and many people in my network, I have been able to scope out a series of hackathon events dedicated to bringing biologists and software developers together to solve technical problems that the biologists face.

The first event will see principal investigators (PIs) and post-docs pitch their problems to selected developers in order to find a problem to solve collaboratively. This will be followed by a few weeks of training and research to discover key questions to answer, and bring the developers up to speed on the basics. Finally, this will be capped off with a weekend hackathon to build tools that solve the challenge posed by the PIs.

Already I have spoken with professors at major UK universities (who have plenty of problems to solve), and I have announced the programme at Hacker News (*https://vimeo.com/137001197*) London. Developers have signed up in droves, and space has been offered at one of the UK's best universities. This will be going forward early next year.

While my wider objective is to create an army of biocoders, I recognize that we must necessarily start small and work our way up. There are many problems that we will discover about this format along the way, so I feel that the series should remain in the pilot phase until great methods of solving those problems are answered. For instance, we need to know what biology topics computer scientists universally have trouble learning about. What are the best analogies to use, or stories to tell, to facilitate their learning process? What troubles will biologists run into when trying to understand the limitations of computer science—because it's not just about putting the tech field in biotech, but also having the biotech field understand tech. We will answer these questions over time, but we need to run as many events as possible.

How You Can Help

The scope of this project is enormous and we need to mobilize as many people as possible in every major tech and biotech hub in the world. I invite *every* company, every lab, and every individual to take part. If we pull off the first few hackathons in different locations around the world, then we can spin this project into a format where *anyone* can run a bio-hackathon and solve important problems in synbio with technology. Going a few steps further, one can imagine that the format could even go beyond software, into hardware or policy and human practices.

If you are a biologist or software developer interested in taking part in the first few events, or if you have any ideas about the future, or if you are otherwise inspired by this idea, please get in touch with me. I'd love to speak with you and find out how we can work together on this project.

Let's make the renaissance in software tools happen. Let's involve the biggest technology community our species has in the most important technology our species is developing.

Edward Perello *is the founder of Desktop Genetics, a London-based software company focused on building CRISPR gene editing software tools. He is a SynBio LEAP fellow and chairs the iGEM software committee. You can reach him at ejlperello@gmail.com.*

Lightning Source UK Ltd.
Milton Keynes UK
UKOW06f2115281015

261579UK00012B/63/P